Donated by

Highlands Cashiers Health Foundation

WORLD OF DRONES

Military DRONES

Hal Marcovitz

San Diego, CA

About the Author

Hal Marcovitz is a former newspaper reporter and columnist who makes his home in Chalfont, Pennsylvania. He has written more than two hundred books for young readers.

© 2021 ReferencePoint Press, Inc.
Printed in the United States

For more information, contact:
ReferencePoint Press, Inc.
PO Box 27779
San Diego, CA 92198
www.ReferencePointPress.com

ALL RIGHTS RESERVED.
No part of this work covered by the copyright hereon may be reproduced or used in any form or by any means—graphic, electronic, or mechanical, including photocopying, recording, taping, web distribution, or information storage retrieval systems—without the written permission of the publisher.

Picture Credits:

Cover: sibsky2016/Shutterstock
6: US Air Force
9: US Air Force
11: Allison Dinner/ZUMA Press/Newscom
15: US Air Force
19: J.M. Eddins Jr./Planetpix/ZUMA Press/Newscom
21: US Air Force
25: Spixel/Shutterstock.com
29: Toru Hanai/Reuters/Newscom
32: Morris Mac Matzen/Reuters/Newscom
35: Darpa/ZUMA Press/Newscom
39: Associated Press
42: Amir Cohen/Reuters/Newscom
45: GH!/Johns Hopkins University/WENN.co /Newscom
49: Bridgeman Images
52: Just Super/Shutterstock.com
55: Cover Images/ZUMA Press/Newscom

LIBRARY OF CONGRESS CATALOGING-IN-PUBLICATION DATA

Names: Marcovitz, Hal, author.
Title: Military drones / by Hal Marcovitz.
Description: San Diego, CA : ReferencePoint Press, Inc., [2021] | Series: World of drones | Includes bibliographical references and index.
Identifiers: LCCN 2019058579 (print) | LCCN 2019058580 (ebook) | ISBN 9781682828335 (library binding) | ISBN 9781682828342 (ebook)
Subjects: LCSH: Drone aircraft--United States--Juvenile literature. | Uninhabited combat aerial vehicles--United States--Juvenile literature.
Classification: LCC UG1242.D7 M338 2021 (print) | LCC UG1242.D7 (ebook) | DDC 623.74/690973--dc23
LC record available at https://lccn.loc.gov/2019058579
LC ebook record available at https://lccn.loc.gov/2019058580

CONTENTS

Introduction   4
Drones: The Future of Warfare

Chapter 1   7
What Is a Military Drone?

Chapter 2   17
Piloting a Drone

Chapter 3   27
Military Drones on Land and Sea

Chapter 4   37
Using Drones for Medical Evacuations

Chapter 5   46
How Drones Are Rewriting the Rules of War

Source Notes   57
For Further Research   60
Index   62

INTRODUCTION

Drones: The Future of Warfare

For three years the terrorist group known as the Islamic State in Iraq and Syria (ISIS) dominated large swaths of territory in the Middle East. ISIS leaders conducted a brutal campaign to enslave the populations of Iraq and Syria, murder its enemies during horrific public displays of torture and execution, and bring a fundamentalist interpretation of the Islamic faith to the people it dominated. In 2017 an international coalition of allied nations, including America, led a military campaign to defeat ISIS. Leaders of the group were captured, killed, or driven into the Syrian and Iraqi deserts, where they were forced into hiding. Yet even after this, the group was still active in parts of Iraq and Syria.

This is why, in November 2018, a US military field commander based in Syria placed a call to Ellsworth Air Force Base in South Dakota. He asked for a drone aircraft (also known as an unmanned aerial vehicle, or UAV) to be deployed over the Syrian town of Baghouz, where an ISIS cell was believed to be located. A drone pilot, sitting behind a bank of computer screens at Ellsworth, directed a US drone known as the Reaper to Baghouz. The drone quickly spotted ISIS soldiers and tracked them to a hideout in Baghouz. The Reaper spent the next two days circling overhead, providing surveillance to military leaders on the ground through the use of sophisticated cameras aboard the drone. Finally, when the Reaper's surveillance made it clear the ISIS soldiers were planning another attack, commanders on the ground relayed an order to the drone pilot to destroy the hideout. Moments later, the Reaper dispatched two deadly missiles. Within seconds, the ISIS outpost was obliterated; all ISIS fighters inside were killed. According to a US Air Force intelligence official, "After that, there weren't any other attacks in the area."[1]

Locate, Identify, and Destroy

The Reaper is the deadliest drone in the US military arsenal. It has a wingspan of 66 feet (20 m) and a top speed of 300 miles per hour (483 kmh). The Reaper can stay aloft for several days at a time and is armed with a devastating air-to-ground missile known as the Hellfire, which can zoom toward an enemy position at about 1,000 miles per hour (1,609 kmh).

The attack on the ISIS hideout by the Reaper illustrates the awesome potential of military drones to locate, identify, and destroy enemies. Most military experts believe drones have the potential for changing the way wars are fought. In future armed conflicts, it may no longer be necessary for squads of troops to risk their lives by swarming into an urban battleground in a foreign city where armed fighters are hidden on rooftops or behind parked cars or lurking beneath darkened windows.

Instead, as shown in Baghouz, drones circling overhead can easily find the enemies and take them out with deadly missile strikes. As Daniel Bruntstetter, a political science professor at the University of California, Irvine, explains, "The three major reasons drones are seen as the future of warfare are: they remove the risk to our soldiers, they make fewer mistakes than other weapons platforms, and technology will continue to improve such that drones become even more precise, efficient, and infallible in the future, thus rendering less precise, efficient and fallible human forms of war obsolete."[2]

obsolete
No longer useful

Raining Down Missiles from Above

Although the United States is believed to have the largest inventory of military drones—the US Army, Navy, and Air Force operate an

The US air force MQ-9 Reaper, pictured here flying over a test range in Nevada, is the deadliest drone in the US military arsenal. The Reaper can stay aloft for several days at a time and is equipped with a devastating air-to-ground missile known as the Hellfire.

estimated ten thousand UAVs—other countries have either developed their own models of military drones or purchased them from the United States or other allies. In fact, according to the Center for the Study of the Drone at Bard College in Annandale-on-Hudson, New York, the militaries of at least ninety countries were believed to be armed with UAVs by 2020. Among those countries are Canada, France, Germany, Great Britain, and Israel.

However, even rebel armies and terrorist groups are known to employ drones. In Yemen in 2019, rebels known as the Houthis used ten drones to attack an oil field in Saudi Arabia in retaliation for military attacks against the Houthis by the Saudis. The Houthis are believed to have obtained their drones from Iran. Yahia Sarie, a spokesperson for the Houthis, said the Saudis can expect more drone strikes. "The only option for the Saudi government is to stop attacking us,"[3] warned Sarie.

The attack on the ISIS fighters in Baghouz and the assault on the Saudi oil fields show how drones have become an integral part of modern warfare. It seems certain, then, that as nations continue to fight nations, and as terrorism remains a very real threat, in the future most of the fighting will not be carried out by soldiers trading gunfire on the battlefield. Instead, remote-controlled aircraft will rain down missiles from above.

CHAPTER 1

What Is a Military Drone?

When drones are described as the future of warfare, the discussion usually centers on aircraft with names like the Reaper and the Predator—hostile names intended to strike fear into the hearts of enemies. They fly tens of thousands of feet overhead and yet are equipped with cameras so keen they can record and transmit the tiniest details below, such as the color of clothes worn by people walking down a street or whether a suspect spotted below is carrying a weapon. They can stay aloft for days at a time, constantly circling silently overhead. Many drones are armed with missiles that can destroy a target seconds after firing. And, perhaps most significantly, they are flown not by onboard pilots but by members of the military sitting behind banks of computer screens thousands of miles away. If a UAV is discovered by the enemy and targeted with a ground-to-air missile, the drone pilot is in no danger.

Drones may represent the future of warfare, but they have also played a role in its past. The first military drone flew during World War I—more than a century ago. Anybody who witnessed that first drone flight on September 14, 1918, would have hardly predicted the evolution of the drone into a fierce weapon of destruction.

A French air force captain, Max Boucher, outfitted a military biplane known as the Voisin with gear that enabled him to control the flight from the ground using a radio transmitter. Boucher steered the plane into the air from an airfield near the French town of Chicheny. The Voisin was in the air for less than an hour. During that time, it traveled about 60 miles (97 km), finally returning to Chicheny for a successful landing.

The Lightning Bug

Although the flight had been a success, the next day Boucher received orders to shut down the drone program. The German army had surrendered (although the official cease-fire was still nearly two months off). Nevertheless, the French military decided it no longer needed new and innovative ways to kill the enemy, and Boucher's drone project was halted. Nonetheless, experiments with drone technology continued elsewhere. In one significant case, the technology was developed not by a team of engineers but rather by an actor.

British film star Reginald Denny had moved to the United States following World War I. When he was not busy in Hollywood, Denny pursued his hobby of flying radio-controlled model aircraft. In fact, Denny soon started designing and building his own model planes; in 1934 he opened a hobby shop to sell his models to other enthusiasts. In 1936 he expanded the hobby shop into a factory to build the models, naming his company Radioplane. In 1939 Radioplane developed the RP-1, a 42 pound (19 kg) monoplane with a 12 foot (3.7 m) wingspan controlled by technicians stationed on the ground. The US Army, preparing for its inevitable entry into World War II, saw the potential of the RP-1 not as a weapon but as a target for practice by ground-based artillery crews and airborne fighter pilots. It ordered fifteen thousand RP-1 drones with the single purpose of shooting them down. That purchase effectively established the drone as an integral tool of the modern military.

Development of military drones continued after World War II ended in 1945. When the United States entered the Vietnam War in 1964, American troops arrived equipped with UAVs. A jet-powered drone known as the Lightning Bug flew surveillance missions over North Vietnam. The Lightning Bug was piloted by a nearby ground-

Because drones are operated by military personnel using computers that can be thousands of miles away, drone pilots are never in danger if an enemy combatant shoots down the craft. This ability to protect soldiers from harm is one reason drones are considered by many experts to be the future of warfare.

based technician. A television camera housed in the nose of the UAV helped ground commanders locate enemy bases and troop positions in the field.

The Lightning Bug did have its weaknesses, though. The drone was launched while it was already airborne, meaning it was dropped from the belly of a larger piloted aircraft. Also, it did not have landing gear. Thus, when it completed its mission, the land-based operator deployed a parachute to enable the drone to float to the ground. Ideally, a nearby US helicopter crew would snag the drone out of the air as it floated downward. In reality, many of the Lightning Bugs crashed into the ground and were lost. In fact, 1,016 Lightning Bugs were deployed over Vietnam. Of those, 544 were lost after crashing or being shot down by enemy fire.

Project Aphrodite

Drone technology did not take a great leap forward during World War II. For example, Project Aphrodite called for equipping bombers with radio control, then filling the planes with explosives. The plan called for a pilot to fly the plane aloft and then parachute out of it once airborne. A technician riding in a nearby human-piloted plane would take over the controls and guide the plane toward a target on the ground. Upon reaching the enemy, the technician would direct the drone to crash into the target, exploding the massive payload of bombs.

However, no Project Aphrodite drones reached enemy targets, largely because the project was developed so late in the war. The planes were designed to be launched from British air bases, flown across the English Channel, and crashed into German positions in France. By 1944, the Germans were in retreat and their troops were no longer within range of the drones. Project Aphrodite was dropped—but not before the program suffered a tragic loss. During a test mission, pilot Joseph Kennedy was killed when the drone he was flying exploded before he had a chance to bail out. Fifteen years after the war, Kennedy's younger brother John was elected president of the United States.

Technology Leaps Forward

By the 1990s, engineers had successfully rid the drones of the technical snags that plagued the Lightning Bugs. The US military's Predator drone, with a 49 foot (15 m) wingspan and a top speed of 135 miles per hour (217 kmh), was first flown in 1995 over the skies of eastern Europe. It helped provide surveillance to peacekeepers quelling Serbian aggression toward the neighboring nation of Bosnia and Herzegovina. By the time of the US invasion of Afghanistan in 2001, the Predator had been armed with Hellfire missiles. The Reaper went into service in 2010. Unlike Predators, which were originally launched for surveillance purposes, Reapers were armed the moment they took to the skies. General T. Michael Moseley, the US Air Force chief of staff during development of the Reaper, noted, "We've moved from using UAVs primarily in intelligence, surveillance and reconnaissance roles . . . to a true hunter-killer."[4] Today the Predator and the Reaper are widely deployed across the skies of the Middle East as well as other troubled regions, helping to protect civilians and troops against terrorists or opposing armies.

Indeed, the technology that enables the Predator and the Reaper to be both reliable and devastating weapons has taken great leaps forward since the days of the Lightning Bug. Perhaps the most significant advancement is the development of the science of radio transmission, enabling pilots to control the drones from outposts located thousands of miles away from the UAV flight paths. In 1898 scientist Nikola Tesla first used radio transmissions to maneuver a toy boat in a pool, laying the groundwork for remotely controlling aircraft movements.

In the early years of UAV development, engineers faced the fact that radio signals tended to die out after a few dozen miles. That is why Boucher could only send radio controls to his plane some 60 miles (97 km) away, or why the Lightning Bugs had to be directed by technicians stationed on the ground near the paths of

The Predator drone was initially flown in 1995 to provide surveillance over eastern Europe. By 2001, the craft had been equipped with Hellfire missiles like the ones shown here being loaded onto an Apache helicopter.

the planes. By the time the Predator was first deployed in 1995, it was able to take advantage of technology that had not been available to Boucher and other early pioneers. The radio signals that guide the Predator, the Reaper, and other modern drones are relayed off of satellites that orbit in space 100,000 feet (30 km) overhead. That is why a drone pilot stationed at an air force base in the Nevada desert can skillfully fly a drone more than 7,000 miles (11,265 km) away over war zones in the Middle East.

> **infrared**
> Electromagnetic energy below the red portion of the visible light spectrum; invisible to the naked eye, rays of infrared light are perceived as heat

Infrared Sensors

The development of orbiting satellites marked a major advancement—but not the only one—shaping the deadly capability of military drones. UAVs are now equipped with onboard technology that has greatly enhanced their abilities to make effective strikes on enemies. Indeed, drones are equipped with numerous cameras and other sensory devices that enable the pilots to have a clear picture of the terrain below the UAV. Today's sensory devices give drone pilots much clearer images than what was available from the 1960s-era television camera that was mounted in the nose of the Lightning Bug.

Another major advancement is the hyperspatial sensor, which is a device that observes objects in the infrared spectrum. The infrared spectrum essentially provides a visual image of an object based on the level of heat it is emitting. Because different objects produce different levels of heat, the hyperspatial sensor is

> **hyperspatial**
> More than a three-dimensional view; a hyperspatial sensor adds an infrared dimension to the visual image

able to determine the identity of an object by the amount of heat it emits. A person emits a different level of heat (normal body temperature is 98.6°F, or 37°C) than, for example, a corner fire hydrant.

That is how the hyperspatial sensor helps tell the pilot that an object on a rooftop is a person and not an inanimate object such as a chimney. Sensors that create images in the infrared spectrum are particularly effective in low-light conditions, such as at night. This is because the human body emits the same amount of heat visible in the infrared spectrum whether it is three o'clock in the

The Black Hornet

Although the US military employs numerous large drones, including the Predator and the Reaper, it has found a use for some very small drones as well. The military's tiniest drone is the Black Hornet. Small enough to fit in one's hand, the UAV weighs just over 1 ounce (32 g). It can be sent aloft with tiny helicopter rotary blades, covering a distance of 1.25 miles (2 km). Troops in the field use the Black Hornet for reconnaissance. The UAV is equipped with a camera that sends images back to a screen carried by the soldier in the field. As they traipse through unfamiliar territory, troops can launch their Black Hornets to observe their immediate terrain, enabling them to locate enemies lurking ahead.

According to technology writer Kyle Mizokami, the

> Black Hornet looks like it could quickly become an infantryman's best friend. Small and quiet, it can flit out and quickly locate enemies. It can peer into the windows of buildings, look behind walls, and detcct ambushes and other traps. In defense, it can keep track of advancing enemies, allowing US troops to more easily repel attacks. In offense it can scout out routes of advance, enabling soldiers to pick out the least dangerous way forward.

Kyle Mizokami, "The US Army Is Putting Tiny 'Black Hornet' Drones in the Hands of Troops," *Popular Mechanics*, May 21, 2019. www.popularmechanics.com.

afternoon or three o'clock in the morning. As Hugh Gusterson, an author and professor of international affairs at the George Washington University in Washington, DC, comments, "A technology that is almost magical gives its owners, who are looking on the scene from high in the sky, a godlike power over life and death."[5]

The Sharp Focus of WAMI

In addition to hyperspatial sensors, modern drones are also armed with other technology that helps them identify enemies below. One such sensor is known as wide-area motion imagery, or WAMI. Basically, the drones are equipped with multiple cameras all scanning the same areas at the same time. The ground-based pilot can flip through the simultaneously produced images, which give an exceptionally sharp focus on the scene below. Camera resolution—essentially the sharpness of an image—is measured in pixels, which are tiny dots that are grouped together to form the image. A camera found in the typical smartphone forms an image using about 12 million pixels. The resolution of cameras employed in WAMI is much sharper: the images are formed using 1.8 billion pixels. Therefore, WAMI enables the pilot to clearly see an object as small as 6 inches (15 cm) even though the drone may be circling several thousand feet overhead.

> **wide-area motion imagery**
> Technology that employs multiple cameras to scan wide fields of view, also enabling the user to identify very tiny objects within that field

Moreover, WAMI is capable of scanning wide areas—even whole cities—at the same time. Therefore, the drone can identify a suspicious person or object miles away, then zoom in on the image, getting a very clear picture of what could be a very small object. WAMI enables the drone operator to record all the images created during the flight, giving the drone operator the power to scroll back to investigate the source of the attack. Technology writer Kyle Mizokami says,

> With WAMI, a drone can point its unblinking eye at the ground below, capturing a medium definition image twice a second. If, for example a car bomb is detonated within the field of view, WAMI will record it. The user can then scroll backward in time, like a DVR recording, and watch the car bomb being driven out of a garage, down the street, and into the location where it will ultimately explode. That sort of information can ultimately lead investigators to the perpetrators.[6]

Shadows, Hawks, and Sentinels

Drones such as the Predator and the Reaper are armed with hyperspatial sensors and WAMI as well as Hellfire missiles, which allows them to spy on the enemy from above and then open fire. They are just two of several UAVs employed by the military. Another drone found in the military's arsenal is the Shadow, which is a small UAV equipped with high-definition cameras employed

The Global Hawk unmanned aerial vehicle is used by the military for surveillance. Because it can fly at an altitude of 7 miles (11 km), it can operate well out of range of ground-to-air missiles.

by the army for surveillance. The Shadow has a length of just 11 feet (3.4 m) and a wingspan of 14 feet (4.3 m). It is launched from a slingshot-like device that can sit atop a vehicle—meaning the army can send it aloft from remote terrain. No long takeoff strip is needed to send the Shadow into the sky.

The Global Hawk is a UAV employed by the air force for surveillance purposes. With a wingspan of 131 feet (40 m) and a speed of 300 miles per hour (483 kmh), the Global Hawk is able to cover more than 14,000 miles (22,531 km) of terrain before it has to be refueled. Moreover, because the Global Hawk is able to fly at an altitude of 7 miles (11 km), it can perform its tasks well out of range of ground-to-air missiles.

The Sentinel drone is also a surveillance UAV, very similar to the Global Hawk. A major difference, though, is that it is not operated by the US military but by the Central Intelligence Agency (CIA), which is America's chief spy agency. It is believed the Sentinel was employed by the CIA to locate terrorist leader Osama bin Laden in 2011. The Sentinel gathered intelligence on the mastermind of the 2001 terrorist attacks on the United States as he hid in a compound in Pakistan. The Sentinel continued circling above, acting as a lookout, as US Navy commandos attacked the compound and killed Bin Laden.

Technology, including WAMI and hyperspatial sensors, has helped make military drones into the fearsome weapons they are today. Moreover, armed with missiles that can travel 1,000 miles per hour (1,609 kmh), UAVs give enemy combatants virtually no chance to escape the destruction that can be rained down upon them from above. Since the era when Boucher first launched a drone in the sky over France, it took some eight decades before the capabilities of drone warfare were fully realized during the US war in Afghanistan. It seems certain that, going into the future, drones will continue to be a significant part of armed conflict.

CHAPTER 2

Piloting a Drone

Military drone pilots do not sit in the cockpits of their aircraft as they soar through the sky tens of thousands of feet overhead. Rather, they sit behind banks of computer screens, with keyboards, joysticks, and throttles within easy reach. Their bases are located thousands of miles from the flight paths of their drones. Communicating with their drones through radio signals, they guide them on their missions through the sky. Using the sophisticated cameras aboard their drones, they record movements on the ground, feeding that intelligence to the commanders in the field. They also may spot enemies and fire missiles at the targets below.

The Skill Set for Piloting Drones

To many observers, the skill set required to fly a military drone closely resembles the skills employed by video gamers. Psychologists at the University of Liverpool in Great Britain wondered whether video gamers would make good drone pilots. In 2017 they invited several traditional military pilots and several avid video gamers, with no experience as aircraft pilots, to take a number of tests. The tests were designed to assess their hand-eye coordination as well as their ability to think quickly and react to changing circumstances. To conduct the tests, the psychologists employed computer simulations of drone missions and asked the participants to fly their virtual UAVs through a number of likely scenarios. Many of the scenarios required the participants to fire missiles at targets on the ground.

After watching the participants in action, and gauging how well they reacted to the simulated drone missions, the Liverpool psychologists concluded that the gamers could do just as well at flying drones as actual pilots. For example, their ability to fly the drones through virtual airspace matched those of pilots who had flown real aircraft. Moreover, psychologist Jacqueline Wheatcroft found that video gamers showed no fear while piloting the simulated drone missions. Because real-life military pilots have often faced terrifying situations while aloft, Wheatcroft says, it is likely that many pilots would bow to their emotions during their drone flights and call off their missions before releasing their missiles. She said such emotional responses did not seem to be a factor that would get in the way of a gamer completing the mission. Wheatcroft concludes, "All we're saying is that [gamers] could have a good skill set for this task."[7]

John Fields agrees. He trains military drone pilots and has often found that trainees who are active video gamers seem particularly well suited to flying real drones. Fields says, "I think a lot of the skills that video gamers have in hand-eye coordination, the ability to multitask and that type of stuff carry through to flying [drone] aircraft systems."[8]

hand-eye coordination

A skill in which a person can carry out tasks that require the simultaneous use of the hands and eyes; essentially, the eyes relay information to the brain, which instructs the hands what to do

Rigorous Training

Drone pilots are not traditional military pilots. To fly most types of aircraft for the US Air Force, Navy, or Marines—including the top-notch supersonic fighter jets—pilots must first be officers, which means they must be college graduates. (US Army pilots do not have to be officers; however, the army does not fly high-performance jets. Most army aircraft are either fixed-wing planes that fly supply missions or helicopters.) Many of the military's best pilots are graduates of the US Air Force Academy in Colorado Springs, Colorado, and the US Naval Academy in Annapolis, Maryland. Prospective marine pilots also attend the naval academy. Both schools are four-year colleges that award bachelor degrees. The two service academies are highly selective: candidates for admission must show excellent academic records as well as other skills the military values, such as leadership and athleticism.

supersonic

Velocity that exceeds the speed of sound, which is 768 miles per hour (1,236 kmh)

A student pilot in San Antonio undergoes training on how to operate a Reaper unmanned aerial vehicle. Because drone operators use keyboards, joysticks, and throttles to operate their crafts, their skills resemble those of video gamers.

After earning their college degrees, the training continues. Prospective pilots must attend and graduate from flight training schools. Programs that train navy, marine, and air force pilots typically take a year. It is a grueling regimen that includes classroom work but also rigorous tests of students' physical skills and powers of concentration. Jet pilots, for example, undergo g-force training in a device known as a centrifuge. As a plane flies higher and faster, the force of gravity exerted against a pilot's body increases. The centrifuge spins a pilot around in circles, artificially increasing the g-force. It is not considered a fun experience; it can be dizzying and painful, and it often makes the pilots ill. "Essentially, you are placed in a chamber that is on the end of a long arm that spins around the room—the faster it spins, the more G-forces you feel on your body," says pilot Jack Stewart. "It feels like weight is pushing down on every part of you."[9] Finally, after four years of college and one year of flight school, the newly minted pilots are awarded their wings and are assigned to duty.

Drone pilots take a much different path to active duty. For starters, they do not need to be officers—enlisted personnel are eligible for the jobs. And since drone pilots do not need to be officers, they do not have to be college graduates. A prospective drone pilot need only have a high school diploma or general equivalency diploma. (For years, the air force had insisted that its drone pilots be officers and, therefore, college graduates, but in 2016 it started accepting enlisted personnel as drone pilots. Air force leaders said they were forced to expand the ranks of drone pilots to enlisted personnel because of a shortage of officers available to fly the thousands of drones operated by the service.)

After entering the service and completing basic training, prospective drone pilots must enroll in a training program. The US

> **centrifuge**
> A device that spins quickly around a central axis, creating intense forces of gravity

Army's program is located at Fort Huachuca in Arizona. The Fort Huachuca training program lasts five months. Unlike flight training programs, in which trainees are put through a number of rigorous mentally and physically demanding exercises, student drone pilots spend most of their training in classrooms and behind computer terminals, where they practice flying drones through computer simulations. Following the completion of the initial course, drone pilots are then assigned to more formal training programs where they learn to control the specific UAVs to which they will be assigned. The second phase of training typically lasts an additional four months.

During training at Holloman Air Force Base in New Mexico, an officer from England's Royal Air Force prepares to spin in a device known as a centrifuge. Centrifuge training is physically grueling. Drone pilots do not need to undergo this type of training.

Boredom Behind the Controls

When the drone pilots finally complete training, they are dispatched to a number of US military bases where they are put into the pilot's seat—although the pilot's seat will, of course, never leave the ground. Still, the drone pilot is flying an aircraft, making it bank left or right, soaring high or diving low, conducting its surveillance duties, and awaiting orders from commanders in the field to open fire. "You do actually fly it," says air force drone pilot James Klein. "You do have a joystick, you have [a] throttle, you have all that, so you do fly the aircraft."[10]

Drone Pilots in Training

The US Air Force trains its drone pilots at Holloman Air Force Base in New Mexico. During training, much of the pilots' time is spent in the classroom. There, instructors discuss scenarios that drone pilots might face, such as how to tell an enemy combatant from an innocent civilian. Eventually, though, the trainees head out to a small trailer where they have the opportunity to fly a Reaper over the New Mexico desert.

Journalist Corey Mead accompanied two pilot trainees, whom he identified as Paul and Justin, during a training session. Paul and Justin flew the drone through various maneuvers over the desert under the watchful eye of their instructor, whom Mead identified as Patrick. Mead writes,

> They crowded inside a dim, chilly, cramped trailer where they sat haloed in the queasy light of 14 computer screens. To an outsider, the next several hours were a mind numbing series of practice runs as Patrick guided Paul and Justin through every possible missile- and bomb-strike [scenario]. To the students, however, this time was not only exciting, but utterly high-stakes. However well they performed in their classrooms, their ability to operate a drone was the only skill that would ultimately matter.

Corey Mead, "A Rare Look Inside the Air Force's Drone Training Classroom," *The Atlantic*, June 4, 2014. www.theatlantic.com.

Klein is assigned to Creech Air Force Base near Las Vegas, Nevada, where he flies a Reaper deployed over the Middle East. The drone is often already in the air when Klein reports for his eight-hour shift. In such cases, he simply takes over the controls from another pilot, and the mission continues.

Newly minted drone pilots who grew up playing video games and expect their duties to resemble the nonstop action found on the screen of a gaming console will soon be disappointed. Although the Reaper is armed with air-to-ground missiles, Klein says he is rarely called on to deploy the weapons. Mostly, he flies reconnaissance missions. Although he has occasionally received orders to fire, Klein says that the Reaper serves in a reconnaissance role—keeping an eye on suspicious activity on the ground—for 99 percent of the time that he is at the controls. Indeed, flying a drone can be a very boring way to spend an eight-hour shift. Former drone pilot Michael Haas comments, "You spend a whole lot of time watching buildings spinning on a screen. Enough to make you crazy."[11]

Ordered to Fire

But, inevitably, the order comes to fire. Brandon Bryant flies drones from a bank of computer consoles located at Creech Air Force Base. Just twenty-one years old, Bryant was assigned to fly a Predator drone over Afghanistan. While on duty one day at Creech, Bryant received orders to focus the drone's cameras on three suspected terrorists below. He viewed the men through the visible spectrum of the cameras on the drone, then switched to the infrared spectrum, which showed him their heat signatures. Each body glowed red on his computer screen. Minutes later he was ordered to fire.

Bryant locked the Predator's lasers onto the men below; these high-intensity beams of light would help guide the Hellfire missile toward its target. Bryant ignited the missile. The Hellfire zoomed away from the drone. Suddenly, Bryant's screen flashed white—an indication that the Hellfire had hit the target and exploded. Bryant describes what he saw next:

> The smoke clears, and there's pieces of the two guys around the crater. And there's this guy over here, and he's missing his right leg above his knee. He's holding it, and he's rolling around, and the blood is squirting out of his leg, and it's hitting the ground, and it's hot. His blood is hot. But when it hits the ground, it starts to cool off; the pool cools fast. It took him a long time to die. I just watched him. I watched him become the same color as the ground he was lying on.[12]

Post-Traumatic Stress Disorder

Many military drone pilots have witnessed similar scenes. And the experience of seeing the devastation caused by their missiles illustrates a significant difference between the job of the drone pilot and the job of a traditional military pilot. After the drone fires its missile, the UAV typically remains in the area, circling overhead. The cameras on the drone enable the pilot to witness the scene after the missile strikes the target, whereas a fighter pilot called on to fire a missile does not usually stick around after it has launched. In fact, by the time the missile hits its target, the fighter—zooming overhead at a speed of hundreds of miles an hour—is already many miles away from the scene. The fighter pilot never sees the destruction he or she has caused below.

Constantly witnessing the horrific scenes caused by their missile strikes has led many drone pilots to suffer from a condition known as post-traumatic stress disorder (PTSD). PTSD is a mental illness caused by witnessing or undergoing terrifying events. It can develop in someone who is a victim of a violent act or in someone who has witnessed or committed a violent act. People who suffer from PTSD often experience flashbacks—in which they see the event occurring over and over again—as well as

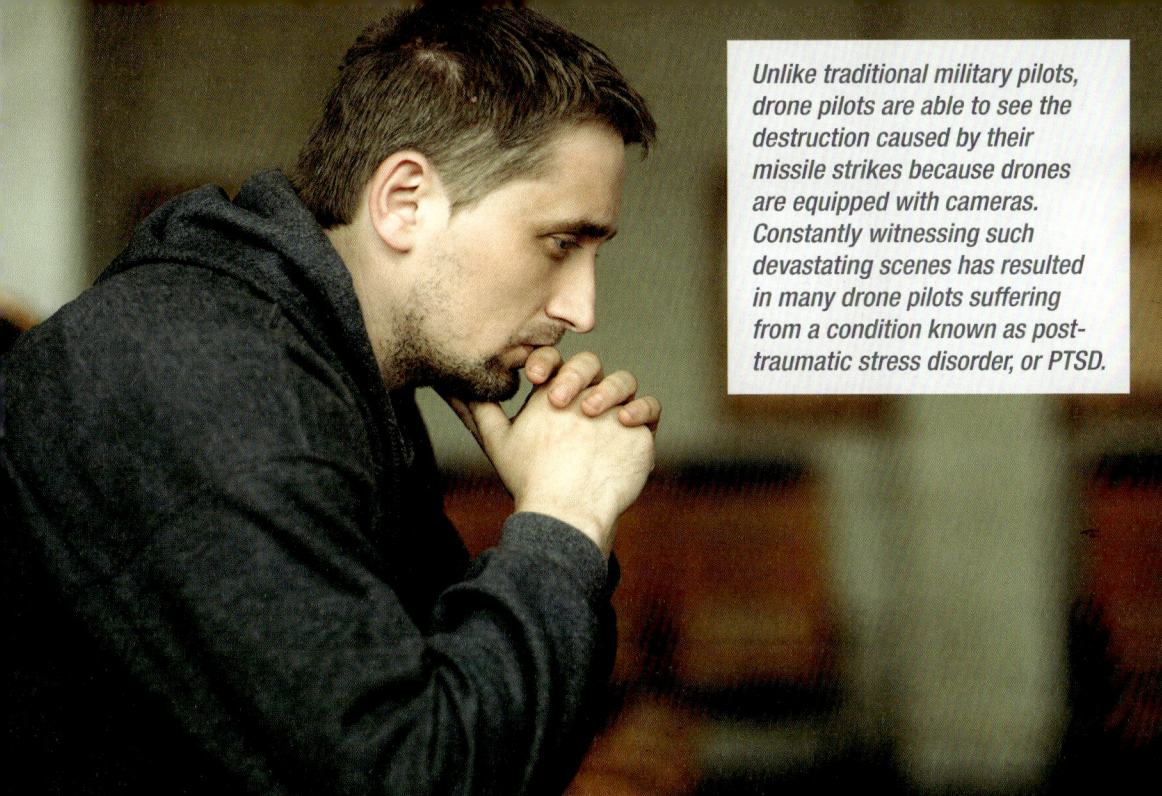

Unlike traditional military pilots, drone pilots are able to see the destruction caused by their missile strikes because drones are equipped with cameras. Constantly witnessing such devastating scenes has resulted in many drone pilots suffering from a condition known as post-traumatic stress disorder, or PTSD.

frequent nightmares and deep anxiety. People who suffer from PTSD often have difficulty functioning in society: they cannot do their jobs, care for loved ones, and otherwise lead normal lives.

Sadly, many war veterans suffer from PTSD after having experienced violent acts on the battlefield. But drone pilots never set foot on the battlefield—after their shifts are completed, they go home to their families. Still, a 2013 study by an agency known as the US Armed Forces Health Surveillance Center assessed the mental health of more than seven hundred military drone pilots and found that 8 percent of the pilots suffered from some degree of PTSD.

A drone pilot named Steven described how his life had changed after witnessing the death, injuries, and destruction he caused by firing missiles at enemy combatants. He said he became numbed to the idea of death. For example, when his grandmother died, Steven said he felt no emotion at all: no sorrow at her loss, no feelings that he would miss her. "You're seeing more death than you [see] normal things in life," he said. "It's still weird

Help for PTSD Sufferers

To help drone pilots endure the mental rigors of their duties, the US Air Force has assigned physicians and psychologists to drone pilot squadrons—enabling the medical professionals to provide early diagnoses and treatments of post-traumatic stress disorder (PTSD) among the pilots. "They are going literally from combat to cul-de-sac in a short drive," says Alan Ogle, an air force psychologist. "Ten minutes, 15 minutes, [they] drive home. They've gone from being eyes [and] head in the fight, and making critical life and death decisions, to then being involved in all the normal . . . responsibilities that we have, where they're a spouse, they're a parent."

Ogle says that in many cases, all the pilots need is a sympathetic ear—someone who can give them moral support and assure them that what they are doing is helping to protect America and its foreign allies. "Sometimes people need to talk about very specific mission details," Ogle explains. "They need to think through what happened in that previous mission, talk through, 'Hey, did I really do what I needed to do, or could I have done something different?'"

Quoted in Sarah McCammon, "The Warfare May Be Remote but the Trauma Is Real," *All Things Considered*, NPR, April 24, 2017. www.npr.org.

taking another life. Distance brings it through a screen, but it's still happening, and it's happening because of you."[13]

Not Like a Video Game at All

The fact that 8 percent of military drone pilots are likely to suffer from PTSD—even though they never come within thousands of miles of an actual battlefield—illustrates how flying a drone is very different from playing a video game. Video gamers may be able to push the right buttons and deftly maneuver through a virtual world of missiles and bombs. When it comes to actually pulling the trigger, killing a person, and watching that person suffer or die, it turns out that piloting a drone is like no video game at all—but a very real part of modern warfare.

CHAPTER 3

Military Drones on Land and Sea

The US Marines are known for their bravery, toughness, and resolve as well as their skills at ground combat. The image of heroic marines landing on shore and assaulting a well-entrenched enemy position is a familiar element of modern culture. In fact, marines have a term for their style of fighting: Marine Air-Ground Task Force (MAGTF), which they pronounce as "MAG-TAF."

MAGTF describes how marines assault the enemy—typically, with a large number of troops deployed on the battlefield by parachuting from above or by rolling into the trouble zone in vehicles, aquatic landing craft, or on foot. That is how marines have engaged in battles for decades, from assaults on Japanese-held islands during World War II to the search-and-destroy missions in the jungles of North Vietnam to street-by-street urban battles that marked the Iraq War during the early 2000s.

But as technology becomes more and more of a factor in modern warfare, many military leaders are acknowledging that MAGTF may not be a part of the marine strategy in the future. According to General Benjamin Watson, "We're no longer going to stick to or take an uncompromising position on the sanctity of the MAGTF."[14]

Watson is head of a program known as the Marine Corps Warfighting Lab. The program is charged with finding ways to employ technology to assist marines when they are called into battle. Many of the projects of the engineers at the Warfighting Lab have focused on the development of drones. In this case, the corps is not looking at airborne drones but rather at land-based machines that are armed with cameras to provide surveillance and guns to destroy enemies before marines arrive at the scene.

Ambling into Battle

One such land-based drone under study by the Warfighting Lab is the Modular Advanced Armed Robotic System, or MAARS. No larger than a golf cart, MAARS is controlled by a marine technician following behind at a safe distance. Directed by the human operator, the MAARS drone ambles through a battlefield on steel tractor treads. Its cameras can pan and tilt, giving the operator trailing behind a 360° view of the terrain on his or her laptop screen. MAARS is also equipped with infrared sensors, identifying enemy combatants by the heat emitted from their bodies.

A high-intensity spotlight is mounted on MAARS. When turned on, it can illuminate a pitch-dark battlefield. MAARS is also armed with two deadly machine guns. Captain James Piniero, a project leader at the Warfighting Lab, says that "[MAARS] allows Marines to understand the threat, understand the environment quicker, and basically see the threat before it sees them."[15]

modular

Part of a larger unit but capable of being used separately

The development of MAARS by the US Marine Corps illustrates how drones are not only changing the way wars are fought in the skies but also on land. In fact, the US armed forces are looking at even bigger and deadlier land-based drones than MAARS. By 2020 the US Army had a number of drones under development that it hopes could take the place of tanks and other armed vehicles on the battlefield.

Members of the US Marines parachute from a plane during a training exercise in Japan. As technology becomes more and more of a factor in modern warfare, many military leaders are acknowledging that traditional marine attack strategies may not be used in the future.

Light, Medium, and Heavy Drones

Three of the army's land-based drones are based on the same concept: a heavily armed vehicle that carries the same firepower as a tank but without the need for a crew to ride inside. The smallest of these weapons is known as the remote combat vehicle (RCV) light. It weighs 10 tons (9 t) and is armed with machine guns as well as rocket launchers capable of knocking out enemy combatants some 2 miles (3.2 km) away. Another land-based drone under study by the army is the RCV medium. This drone weighs 12 tons (11 t) and includes all of the remote-controlled features of the light model; however, instead of a rocket launcher, it is armed with a cannon that fires shells capable of striking targets 2 miles (3.2 km) distant. Finally, the RCV heavy is an even larger version, weighing 20 tons (18 t). The firepower aboard the RCV heavy includes a cannon with a range of 2.5 miles (4 km).

The RCV heavy sounds as though it is a massive piece of armament but,

remote combat vehicle

A land-based drone armed with machine guns, cannons, or rocket launchers

Learning a New Way to Fight

The US Air Force and Navy have been flying airborne drone missions since the 1990s, meaning these branches of the armed services have nearly three decades of experience in training drone pilots and seeing how they react under pressure. But land-based drones are still very much in the experimental phase. The US Army and Marines are just getting started in training land-based drone operators and are therefore unsure of how well soldiers in the field will be able to control the machines.

Unlike the pilots of Predators and Reapers, the drone operators on land will not be sitting in remote stations thousands of miles away, focusing on banks of computer screens. Rather, they will be following at a safe distance behind the land-based drones on foot or in vehicles, with little more than laptops to use in guiding the drones. "This is not how we're used to fighting," says David Centeno Jr., an official in a US Army department charged with developing the drones. "We're asking a lot. We're putting a lot of sensors, putting a lot of data in the hands of soldiers. We want to see how that impacts them. We want to see how it degrades or increases their performance."

Quoted in Stephanie Mlot, "US Army to Test Remote-Controlled Combat Vehicles," Geek.com, July 15, 2019. www.geek.com.

---

in reality, it is still smaller than the M-1 Abrams tank, which is a standard weapon in the army's arsenal. The M-1 Abrams weighs as much as 70 tons (64 t) and is designed to carry a crew of four soldiers. Heavy armor plating is needed to protect the four crew members from enemy fire. Because the RCV drones do not have to protect human occupants, they can be smaller and lighter.

Army combat drones like these might not be as nimble as soldiers, but they can reduce US casualties and still be just as deadly against enemy combatants. "[The army] is looking at . . . combat vehicles that could be much smaller, cheaper and more expendable than its manned vehicles—five, 10 or 20 tons—allowing them to play very different tactical roles," says journalist Sydney Freedberg Jr., who writes about military-related topics. "Yes, human troops can also carry [rocket launchers], and trained infantry will probably be better at hiding than a [drone], even a relatively small one. . . . But once those humans open fire, they will be targeted, and not all of them will make it out."[16]

Seaborne Drones

While the US Army and Marines pursue the development of land-based drones, the US Navy has moved in a different direction. In addition to its existing fleet of airborne drones, the navy sees a future in drones that operate on and below the surface of the ocean.

The navy is pursuing the deployment of seaborne drones for the same reason it uses airborne drones: to keep human crews out of harm's way. Even in times of peace, the ocean can be a dangerous place. Crews often have to endure rough seas and turbulent storms. "In wartime as in peacetime, the sea itself is a deadly adversary of those who travel on it or under it. Even in peacetime, hazards, in the form of strong winds, rough seas, and hidden reefs abound while shipwreck and drowning are an ever present danger," write Robert Sparrow, a professor of philosophy at the University of Melbourne in Australia, and George R. Lucas Jr., a professor at the US Naval Academy. The dangers are compounded by war. "In wartime, seafarers who are forced to abandon ship after an enemy attack may find themselves facing nearly certain doom alone in freezing waters or floating in a small life raft thousands of miles from land."[17]

The technology for seaborne drones is not as far along as for airborne drones. Whereas airborne drones can be piloted from thousands of miles away, seaborne drones cannot. At present, human crews stationed aboard navy ships are still needed to pilot the drones. Radio waves do not travel far underwater, tending to die out after a few miles. Therefore, drone pilots cannot take advantage of satellites to relay signals to the drones traveling underwater. It means crew members must follow nearby, relaying orders to the drones. These crew members must still go to the battle zone—or, at least, to a location near the battle zone—in order for the drones to carry out their missions.

Disarming Mines

Still, seaborne drones are performing other important tasks that keep military personnel out of danger. One of those tasks is the disarming of mines. A mine—essentially a bomb tethered to the ocean floor—can blow a hole in the hull of a ship if it comes into contact with the vessel. Enemies typically mine harbors and other offshore points to guard against an invasion, such as US Marines storming a beach.

Dating back to World War II, the responsibility for finding and removing mines has often fallen on the shoulders of navy scuba divers. But it has always been an extremely hazardous duty. Indeed, safely defusing explosive devices found on urban streets or in remote jungles is very dangerous. Specialists in bomb disposal are often clad in protective armor to shield them in the event the bomb explodes while they work to disarm the device. But

A mine diver demonstrates the disposal of an ocean-borne mine. People have been performing this extremely hazardous work for decades, but today submersible drones are employed instead, thereby eliminating the risk to human divers.

navy divers can wear no such protective armor over their diving gear. Moreover, they are forced to work below the surface of the ocean while often fighting against strong or swirling currents, further complicating what may be intricate bomb-defusing work. Likewise, they are limited to working for only as long as they have air supplies in their tanks—typically no more than an hour. "Divers have to come up every so often. They can only dive for so long and can only swim so fast and cover so much,"[18] says Captain Tim Schnoor, deputy program manager at the Office of Naval Research, a US Navy department studying the use of drones for mine clearance duties.

Such perils are the reason the navy is actively pursuing the development of submersible drones known as the Remote Environmental Monitoring Units (REMUS). Resembling torpedoes, REMUS drones were first deployed for surveillance purposes only, scouting out enemy waters before manned vessels arrived in the battle zones. But the navy also found REMUS drones to be effective for finding and disabling mines. The drones were first used for mine clearance duties during the Iraq War in 2003, when divers discovered mines blocking the Khor Abd Allah, a waterway that ships use to enter Iraqi rivers from the Persian Gulf. In order for navy vessels to gain access to the Khor Abd Allah, REMUS drones were dispatched into the Khor Abd Allah and were able to find numerous mines.

Once a REMUS drone located a mine, it planted a small explosive charge on the device and then sped away. When the drone was out of range, the human pilot detonated the explosive, safely blowing up the mine. "REMUS makes all the difference in these missions, which are dangerous for human divers,"[19] says Arthur Holland Michel, director of the Center for the Study of the

Remote Environmental Monitoring Units

Oceangoing drones used for underwater surveillance and for locating and disarming undersea mines

Drone at Bard College. By 2020 the navy had made numerous improvements in REMUS technology. The REMUS 600 drone, for example, is able to remain submerged for three days while it travels a range of some 280 miles (451 km).

Surface Drones

Although submersible drones such as REMUS have proved very effective in finding and disarming mines, the US military is pursuing research into drones that will travel on the surface as well. One such drone is the Sea Hunter, which is an experimental

Russia's Land-Based Drone Program

Although the US military has conducted extensive research into land-based fighting drones, the machines have yet to be deployed in the field on more than an experimental basis. Other countries are also developing land-based drones and have already sent them into battle. Russia, for example, deployed drones to the Middle Eastern nation of Syria in 2018 to help the Syrian government gain control over territory held by insurgents.

Known as the Uran-9, the land-based drone measures 16 feet (5 m) in length and is armed with two missile launchers capable of knocking out enemy tanks. The drone is also armed with a machine gun for land-based targets as well as a cannon capable of downing low-flying aircraft. Although the Russian army rolled out the Uran-9 to face the Syrian insurgents, Russian military officials acknowledged they were disappointed by the performance of the drone. In fact, the Uran-9 generally failed to respond to remote commands whenever it was deployed. As Russian military research officer Andrei Anisimov explains, "Modern Russian combat unmanned ground vehicles are not able to perform the assigned tasks in the classical types of combat operations." He estimates it will be a decade or more before Russia's land-based drones are ready for combat.

Quoted in Sebastien Roblin, "Russia's Uran-9 Robot Tank Went to War in Syria (It Didn't Go Very Well)," *The Buzz* (blog), *National Interest,* January 6, 2019. https://nationalinterest.org.

The Sea Hunter is an experimental drone that travels on the surface of water. Although it is currently used for surveillance and mine hunting, it may one day be armed with weapons that will enable it to be sent into battle.

130-foot-long (40 m) surface vessel that has undergone testing since 2018. As with the REMUS drones, the Sea Hunter is used for surveillance, as well as finding and disarming mines. However, the Sea Hunter is an improvement on the undersea REMUS drones in that its range is much larger—the craft is capable of traveling some 10,000 miles (16,093 km) before it needs to be refueled.

Currently, the Sea Hunter's duties are restricted to surveillance and mine hunting, but navy officials foresee the day when vessels like the Sea Hunter will be armed with cannons, rocket launchers, and torpedoes and sent into battle. And the reason for arming ships like the Sea Hunter is obvious: no human sailors will be in danger during a battle if no human sailors are aboard the fighting ships. In fact, in 2019 the navy announced its intention to expand the Sea Hunter concept, calling on defense contractors to design vessels as large as 300 feet (91m) in length. The navy refers to these vessels as large unmanned surface vehicles (LUSVs) and has indicated an interest in adding as many as ten LUSVs to its fleet by 2025. "The large unmanned ships will be generally unarmed, but with the ability to accept payloads of anti-ship missiles and land-attack cruise missiles," reports the magazine *Popular Mechanics*. "In that sense they will be an extension of the Navy's fleet of destroyers and cruisers, acting as floating magazines to bolster the fleet with more weapons."[20]

In the near future, with armed drones available not only in the sky but also on land and sea, conflicts between nations will likely be fought by machines rather than by humans. No longer will armed services like the US Marine Corps need to rely on MAGTF tactics to defeat the enemy. The drones will take on that duty.

CHAPTER 4

Using Drones for Medical Evacuations

The Predator, the Reaper, and other airborne drones zoom through the sky at speeds as fast as 300 miles per hour (483 kmh). They can watch enemy combatants while circling miles above, then unleash Hellfire missiles that can obliterate enemy positions within seconds. But the military has found a need for drones that move quite a bit more slowly while carrying no weapons. Unlike most UAVs, these drones are designed not as jets but as helicopters. Rotary blades lift the drones into the air and enable them to descend on a vertical path. The military calls these UAVs vertical takeoff and landing (VTOL) drones.

The drones are needed to transport wounded soldiers from the battlefield to safety, where they can receive medical treatment for their injuries. For decades, helicopters—known as medevac units—have performed that duty. They are flown by pilots and carry crews of military members trained as emergency medical technicians. But finding wounded soldiers in the field and transporting them to safety has always been filled with danger. The helicopter often has to make its way through enemy fire to reach the wounded soldier on the battlefield, putting the medevac helicopter and its crew in danger.

By 2020, the US Army was exploring use of VTOLs that can make their way through hostile airspace to find wounded soldiers. Under this concept, the drone would be controlled by a ground-based pilot entrenched near the battle zone. According to technology writer David Axe, medevac is

> one of the most dangerous missions on the modern battlefield—and one of the most important. Crews flying big, vulnerable and sometimes unarmed helicopters brave gunfire, bad weather and rugged terrain to snatch wounded troops from a firefight or the scene of a bomb blast.
>
> Medical evacuation crews are some of the gutsiest people around. . . . Now the Army has latched onto a possible solution: replace the human copter crews with VTOL drones.[21]

Roots in Twentieth-Century Warfare

Medevac helicopters have long been used by civilian hospitals in the United States and other countries. Many hospitals feature helicopter landing pads on their grounds, enabling civilian-operated medevac helicopters to quickly ferry victims of automobile accidents or other incidents to emergency rooms.

The use of medevac helicopters to transport accident victims to hospitals has its roots in twentieth-century warfare. The US military services first started using helicopters to evacuate wounded soldiers to safety near the end of World War II. Later, medevac helicopters were used widely in the Korean and Vietnam Wars. Many American soldiers owe their lives to medevac teams, whose duty has always been hazardous. According to a report by the American Homefront Project, a group that provides news reports on military life, medevac helicopters were shot down at three times the rate as assault helicopters during the Vietnam War.

medevac

An abbreviation for *medical evacuation*, it is typically carried out by a civilian or military helicopter to fly victims to hospitals

Medevac helicopters like the one shown here in Afghanistan in 2009 rescue wounded soldiers from the battlefield. In the future, such rescues may be performed by remotely controlled drones.

The plight of medevac missions has not improved since the Vietnam War. In 2004 a medevac helicopter carrying wounded soldiers was shot down during the Iraq War. All nine American crew members and wounded soldiers on board were killed. In the intervening years, terrorists and rebel armies have obtained more sophisticated weapons, such as mobile rocket launchers. Given this escalation, US military planners have concluded that human-piloted medevac helicopters may no longer be a viable method of rescuing wounded soldiers. "A drone medical evacuation aircraft is a logical response to these concerns," says technology writer Joseph Trevithick. "Downed pilots or wounded troops effectively trapped behind enemy lines are exactly the kind of situations where an unmanned rescuer, less conspicuous and a smaller target than a large chopper, might be especially useful."[22]

conspicuous

Easily visible to others

Teams of Drones Used for Medevac

The US Army is exploring the concept of employing existing land-based drones for medevac purposes rather than developing new drones specifically to evacuate the wounded. The plan calls for reprogramming existing drones to do the tasks necessary to extract wounded soldiers from battle.

For years, the army has employed a drone known as the Advanced Explosive Ordnance Disposal Robotics System. This small device, which is about the size of a shopping cart, rolls into battle on tractor treads to identify and disarm bombs. (Many city police departments use similar drones.) In such a scenario, several disposal robotics drones would be dispatched to the scene and, working in tandem, provide a litter to carry a wounded soldier to safety.

According to technology writer Michael Peck,

A single [drone] can't do all the tasks required for medevac. . . . However, Army acquisition officials maintain that a group of coordinated small mobile [drones] could team up to collaboratively perform the required tasks of locating, assessing and extracting a casualty, with a human in the loop for supervision and high-level commands. The Army is looking to develop a casualty extraction capability using unmanned air or ground vehicles, perhaps equipped with manipulators or grippers, that could assemble into [a] team as necessary.

Michael Peck, "Army Seeks Swarms of Mini-drones for Medevac," *Defense Systems*, September 13, 2017. https://defensesystems.com.

A Small Target for Enemy Gunners

The drone the army hopes to employ would find the wounded soldier, land in the battlefield, and retrieve the patient, sometimes with help from other soldiers on the ground. One drone under study for this duty is the DP-14 Hawk. This UAV is 13.5 feet (4 m) in length and weighs just 430 pounds (195 kg). The Hawk features twin rotors and can attain a speed of 120 miles per hour (193 kmh). The "cargo" space inside the Hawk is just a mere 2 feet

(61 cm) in width, meaning it is capable of holding just a single wounded soldier. The craft can stay airborne for 2.5 hours. By 2020, the DP-14 Hawk was still very much in the test phase. No Hawk had yet to make a flight with a human passenger aboard.

Still, what the military finds particularly attractive about the DP-14 Hawk is its small size. Under current battle conditions, most medevac missions are flown by the Black Hawk helicopter. The Black Hawk, which is also used for combat assaults, is about 65 feet (20 m) long and can carry eleven people. The size of the helicopter means that, in addition to providing a sizable target for enemy gunners, the Black Hawk cannot touch down in tiny ravines, crevices, or other tight places where a wounded soldier may have taken shelter. "The smaller drones could touch down in relatively tight areas in mountainous, forested or jungle environments where larger medical evacuation helicopters, such as the Black Hawk, would never be able to go,"[23] says Trevithick.

> **rotor**
> The blade that spins atop an aircraft, particularly a helicopter, that enables the craft to ascend vertically

Israel's Medevac Drone

The US Army is not the only armed service looking at medevac drones. The Israeli army has expressed a deep interest in finding methods to extract wounded soldiers from battle zones. For decades, Israel has been engaged in a series of conflicts with hostile neighbors. To fly wounded soldiers out of hostile border regions, the Israeli military is looking very closely at a drone known as the Cormorant (named for a species of aquatic birds).

As with the DP-14 Hawk, the Cormorant is capable of ascending and descending vertically. But the Cormorant will not rely on rotary helicopter blades whirling overhead. Instead, the Cormorant is powered forward by two propellers mounted closely along each side of the UAV. These are known as

The Cormorant (pictured) is Israel's version of a medevac drone. One unique feature of the craft is that it is equipped with devices that monitor the patient's vital signs, such as heart rate and blood pressure, during transport.

ducted fan propellers because the propellers are encased in wide cylinders, or ducts. Finally, two large rotating blades, each 6 feet (1.8 m) in diameter, are located beneath the aircraft, providing it with lift.

The dimensions of the Cormorant are a bit larger than the DP-14 Hawk. The aircraft is 20 feet (6 m) in length and 11 feet (3 m) in width, large enough for the Cormorant to carry two wounded soldiers. The Cormorant also is a bit slower than the DP-14 Hawk, with a top speed of 115 miles per hour (185 kmh), but it can stay aloft for four hours.

The Flight of the Cormorant

As with the DP-14 Hawk, Israeli military officials have not yet authorized the aircraft to be used in actual combat. A 2018 test, though, involved military personnel loading a mannequin into the Cormorant, which then lifted off and successfully transported the mannequin to a remote location.

Technology writer Jack Stewart describes the 2018 test flight of the Cormorant, which carried the mannequin patient aloft:

> Five men in white overalls lifted the stretcher off the ground, one of them taking care to lay a clear plastic [intravenous] bag that's connected to the patient onto his stomach. They marched him toward what looks like a black inflatable dinghy on small wheels. . . . The stretcher was loaded in through a hatch on the side, and then the men stood back. . . .

Flying In Medical Supplies

Although the deployment of drones to evacuate wounded soldiers from the battlefield is still very much in the future, the use of UAVs to deliver drugs, blood supplies for transfusions, and other medical supplies into battle zones may occur very soon. US military researchers have focused on using small fixed-wing and rotor-blade drones—similar to drones available to hobbyists—to ferry supplies to emergency medical technicians in the field. The concept is similar to the use of drones for commercial delivery purposes, which some national retailers are known to be studying.

In 2018 US military leaders contracted with a San Francisco, California, company known as Zipline to test the delivery capabilities of drones. Zipline manufactures a small fixed-wing drone with a wingspan of 7 feet (2 m) and the capacity to carry 4.5 pounds (2 kg) of cargo. For the test, Zipline dispatched eight drones on 227 medical supply delivery missions to an undisclosed area of the United States. "It fills a gap in a capability we don't have, and that's to rapidly resupply life-saving medical supplies," says Lieutenant Shane Kim of the US Marines, a member of the team studying the use of drones to deliver medical supplies.

Quoted in Patrick Tucker, "US Marines Try Using Drones to Bring Blood to Battle," *Defense One*, October 22, 2019. www.defenseone.com.

The rescue of the mannequin was actually the second half of the demonstration the Cormorant put on at a remote airfield in northern Israel. . . . First, it took off with a full cargo of military supplies, slightly wobbly at first. It leveled out as it flew a large loop over a green field, then descended vertically, alighting on the grass. The overall-clad squad offloaded the cargo and put the patient in its place, then watched it take off again.[24]

Robo Sally

One feature of the Cormorant that appeals to medical professionals is that the aircraft is equipped with devices that monitor the patient's vital signs during transport. For example, equipment aboard the Cormorant can monitor the patient's heartbeat and blood pressure. Additionally, television cameras aboard the Cormorant will allow doctors to actually see the patient and determine the nature and severity of the wounds ahead of time.

As of 2020, neither the Cormorant nor the DP-14 Hawk had carried human passengers—even in test flights. The slow development of VTOL drones illustrates the amount of work left to do before commanders trust the drones to ferry their wounded out of battle zones. Moreover, what neither the Cormorant nor the DP-14 Hawk can do is actually evacuate a wounded soldier without human assistance. Both drones can be flown into a combat zone and landed within a few feet of wounded soldiers. If possible, the soldiers would board the drone on their own. Others, however, would require help from their fellow soldiers. Of course, by helping the wounded soldier board a drone, those soldiers may expose themselves to enemy fire. At this point, medevac drones under development feature no robotic mechanisms capable of lifting wounded soldiers aboard the VTOLs.

Yet if a drone dubbed Robo Sally ever sees action, that problem may go away as well. Robo Sally is a ground-based drone designed to roll into a battle zone on wheels. Equipped with dexterous mechanical arms and fingers, Robo Sally is able to lift up a wounded soldier and carry him or her to safety or perhaps deliver the soldier to a Cormorant or DP-14 Hawk, which would then fly the soldier out of the battle zone.

Robo Sally (pictured) is a ground-based drone designed to roll into a battle zone on wheels. Researchers are exploring fitting drones with mechanical arms and fingers dexterous enough to perform delicate tasks like disarming bombs.

Potential to Save Lives

About the size of a lawn tractor, Robo Sally has been under development at Johns Hopkins University's Applied Physics Laboratory in Baltimore, Maryland, since 2013. And Robo Sally is every bit the drone, being operated by a technician located outside the battle zone who can see the environment around the drone through cameras mounted aboard the device. In addition to rescuing wounded soldiers, military planners believe Robo Sally can be effective in performing other delicate and dangerous tasks, such as disarming bombs.

Technology writer Ford Burkhart described a demonstration of Robo Sally's capabilities: "One engineer stood behind Sally and moved his hands and arms, sending signals over wires to Sally. Sally perfectly replicated those motions, including reaching out and shaking hands, as her hand sensor gave exactly the right firmness. In the field, it could all be done remotely, untethered, from up to a half mile away."[25]

As of 2020, Robo Sally was still under development at Johns Hopkins University. As with the Cormorant and the DP-14 Hawk, the drone is not ready for use on the battlefield. If all these drones do make it into the field of battle, they have the potential for saving the lives of many soldiers.

Chapter 5

How Drones Are Rewriting the Rules of War

In the modern age, certain rules of war—largely unwritten—have long been accepted by all combatants. One of those rules protects prisoners of war. If troops in the field encounter enemy soldiers, and those enemy soldiers elect to surrender, they drop their weapons and agree to be taken into custody. The vanquished troops are then held in camps for the duration of the war. They are provided with food, shelter, and medical care. When the war ends, the captured troops are released and sent home.

Taking No Prisoners

But drones like the Reaper and the Predator do not take prisoners. They locate the enemy and then, responding to commands relayed from thousands of miles away, unleash their Hellfire missiles. No drone employed by any military on Earth has yet to detect an enemy position, circle overhead, and broadcast a warning to the combatants below advising them to either surrender or be obliterated by a missile strike.

Therefore, many experts believe that the use of drones has rewritten the rules of war that have been observed by nations for centuries. University of North Carolina history professor F.S. Naiden comments, "If drones are used to the exclusion of short-range weapons, American forces will be unable to take or interrogate prisoners, accept surrenders, and occupy positions. These practical advantages are not to be despised. Some terrorists surrender . . . some provide intelligence. Yet a drone cannot communicate with the enemy."[26]

Moreover, Naiden says, in past wars one side inevitably comes to the realization that it cannot win. Either the enemy's army has gained an overwhelming strategic advantage or its own resources have been depleted. Its arms and ammunition have been spent, its armies lack enough soldiers to carry on the fight, or the home front lacks food and other essentials. At that point, Naiden says, the enemy's leaders usually sue for peace. The two sides meet and negotiate a treaty. For example, as part of the peace process when World War II ended in 1945, the Allies established a plan to rebuild war-torn Europe and fashion Germany into a democracy. Today, Germany is a free and vibrant democracy.

But drone warfare may take that option away as well. It is within the power of a drone to locate the headquarters of the enemy government and, by releasing its Hellfire missiles, completely obliterate the government's leaders. This action would leave no one alive to negotiate a surrender. By wiping out a nation's leaders with one lethal missile strike, Naiden says, drone warfare might actually serve to prolong the conflict. Naiden explains that "once an enemy's leaders are dead, drones will not be able to capture, interrogate, or parley with survivors. And without any leaders to represent them, the remaining forces of the enemy may not be willing or able to surrender. The elimination of whole strata of leaders may not bring the conflict to a close."[27]

strata
A series of layers

War Without End

In some cases, wars have been brought to a close not by the military or through negotiations but rather by public outcry over the seemingly endless battles and the human toll of the conflict. That is how America's involvement in the Vietnam War was brought to an end. Throughout the duration of America's participation in the conflict, from 1964 to 1973, some fifty-eight thousand US troops lost their lives, and tens of thousands of others returned from the jungles of Southeast Asia both physically and emotionally scarred.

Many opponents of drone warfare warn that there is a lesson to be learned from the Vietnam War. During the 1960s, as the conflict in Vietnam escalated, unrest rose on American streets as demonstrators voiced their opposition to the war. Young men of draft age publicly burned their draft cards, declaring they would rather go to prison or flee the United States than fight in Vietnam. Movie stars, rock musicians, and other celebrities denounced the war. Perhaps a major turning point occurred in 1968 when popular television news anchor Walter Cronkite declared on his evening broadcast, "It seems now more certain than ever that the bloody experience of Vietnam is to end in a stalemate. . . . It is increasingly clear to this reporter that the only rational way out then will be to negotiate, not as victors, but as an honorable people who lived up to their pledge to defend democracy, and did the best they could."[28]

In fact, it took five more years, but US political leaders got the message. They negotiated America's withdrawal from the fighting, thus ending the long conflict that had cost so many lives. Had it not been for the unrest at home—sparked by the continued loss of American lives—the war in Vietnam could have dragged on even longer.

Arguably, though, if American troops had never suffered such a staggering death toll there never would have been an antiwar movement back home. And without that homegrown antiwar movement, American politicians would never have felt pressure to end the war. "What had started as . . . a protest of a few grew into a torrent of a vast and representative majority—probably the largest

draft

The US government's system for compelling young men to serve in the military

Demonstrators in 1972 protest the war in Vietnam, which killed fifty-eight thousand Americans. Today, many antiwar activists fear that drones will ensure that American troops are never in danger, so American political leaders will never feel pressure to end wars.

and most effective antiwar movement in history,"[29] writes Todd Gitlin, who was a peace activist during the 1960s.

And so, amid the widespread use of drones in twenty-first-century combat, many antiwar activists fear that if American troops are never in danger, then American political leaders will never feel pressure to end wars. They worry that drone warfare could conceivably continue without end. Such conflicts are likely to cause great devastation to civilians living in countries that are continually under siege from pilotless drones circling above. But in the countries that launched the drones, citizens may have little idea what is going on in those foreign lands. "Drones never come home in coffins," says Amy Zegart, a political science professor at Stanford University in California. "They do not have grieving families. There will be no . . . moments in which video footage showing the death and desecration of American troops leads to anguished demands that the president do something. Because drones pose no risk to the warfighter, they remove a key emotional element that influences domestic politics."[30]

Americans Support UAV Warfare

In fact, there is already evidence to suggest that few American citizens would feel compelled to pressure political leaders to end wars that are fought by drones. Recent national polls suggest that a majority of Americans are on board with sending UAVs to do the fighting that would otherwise be left up to human service members. Over the past few years, national polling agencies have asked Americans whether they support UAV warfare. Each poll has reflected strong support at home for the use of military drones.

For example, a 2015 poll by the Pew Research Center in Washington, DC, showed 60 percent of Americans supporting drone warfare over the use of American soldiers to do the fighting. Likewise, a 2016 poll by the Center for a New American Security, a group that studies national security issues and is also based in Washington, DC, showed 60 percent support among Americans for using drones to carry out combat missions. Polling results suggest that Americans favor drone warfare for the obvious reason that deploying UAVs to do the fighting means no American soldiers need to risk their lives in battle. As Jacquelyn Schneider and Julia Macdonald, researchers for the Center for a New American Security, explain, "The most common assumption about public approval for drone strikes is that the public significantly prefers drones over their manned counterparts because the use of UAVs mitigates risk to US military personnel."[31]

As drone technology moves forward, it is likely that military personnel may have even less engagement with the enemy than they do now when drones are dispatched to do the fighting. Research is already under way to adapt artificial intelligence, or AI, for use in the operation of military drones. Still in the developmental

artificial intelligence

Software that enables a computer to react to a changing environment, sending instructions to a machine it controls much the same way as the human brain would react in the same environment

Undercounting Civilian Deaths

A reason for the American public's support for military drones may lie in the fact that the US military rarely acknowledges civilian deaths caused by drone strikes. A 2012 report by the Human Rights Clinic, a department of Columbia Law School in New York City, found the military routinely underreports the number of civilians killed in drone strikes. The Human Rights Clinic looked at media reports of drone strikes against suspected terrorists in Pakistan in 2011 and found the military acknowledging the deaths of no more than 9 civilians. The group looked at statistics compiled in press reports and concluded that as many as 155 civilians were killed in Pakistan that year. According to the Human Rights Clinic report, "The public has no information on how and whether the US tracks and investigates potential civilian deaths."

Since the report was issued, drone strikes killing civilians have continued. In 2019 a drone killed five civilians riding in a car in Afghanistan, Afghan government officials told reporters. After the strike, US military spokesman Colonel Sonny Leggett said only, "We are aware of the allegations of civilian casualties and working with local authorities to determine the veracity of these claims."

Human Rights Clinic, *Counting Drone Strike Deaths*. New York: Columbia Law School, 2012, p. 6. https://web.law.columbia.edu.

Quoted in Farooq Jan Mangal and Fahim Abed, "US Drone Killed Afghan Civilians, Officials Say," *New York Times*, December 1, 2019. www.nytimes.com.

phase, AI features software that enables a computer to diagnose and interpret data and adapt its decision-making to a changing environment much the same way the human brain performs the same functions.

Applying Artificial Intelligence to Drones

AI technology is currently being developed and tested for use in self-driving cars. Within the next decade or so, it is expected that an AI-equipped vehicle will be able to drive itself—the onboard computer will negotiate turns, accelerate, brake, observe traffic signals, and get the passenger safely and promptly to his or her destination. A number of onboard sensors provide information to the computer, such as the locations of nearby passing cars and pedestrians. That information is fed to the computer, which then employs AI to enable the car to react to the surrounding environment.

Similar technology would be applied to UAVs. This means that when AI is employed in drone warfare, it may no longer be necessary for a drone pilot to sit behind a computer terminal at a military base thousands of miles removed from the conflict, directing the flight of the drone and releasing its missiles. Instead, a computer housed aboard the drone will fly the UAV, locate the enemy, select the target, and release the missiles, thereby taking

This rendering of a self-driving car shows the numerous onboard sensors that feed information into a computer, which in turn uses artificial intelligence to enable the car to react to the surrounding environment. Drones may someday be equipped with the same technology, eliminating the need for human operators.

Lockheed Martin's Drone Races

Lockheed Martin, a major aerospace corporation that supplies many of the drones flown by the US military, is developing drones piloted by artificial intelligence (AI) technology under its AlphaPilot program. "Right now, autonomous drones are a thing you'd only find in labs, being pioneered by a small, niche audience," says Keith Lynn, Lockheed Martin's program manager for AlphaPilot.

To encourage development of AI-piloted drones, Lockheed Martin sponsors an annual series of competitions pitting its AlphaPilot drones against drones developed by other companies as well as university engineering departments and individual entrepreneurs. The winner of the annual series of four competitions is awarded $1 million by Lockheed Martin. The first series of competitions, staged in 2019, initially attracted interest from four hundred developers of AI-piloted drones. When the competition began, however, it drew only nine entrants. Despite widespread interest in AI-piloted drones, many developers do not believe their technology is ready for competition. The AI-piloted drones that competed in the 2019 series hardly resembled Reapers and Predators. The machines were tiny, resembling drones available for sale to hobbyists. And they were flown through courses laid out inside gymnasiums, minimizing dangers to passersby. Nevertheless, the drones made use of onboard sensory equipment to find their way through a series of obstacles on their own, without the need for guidance by pilots.

Quoted in Jake Swearingen, "AI Is Flying Drones (Very, Very Slowly)," *New York Times*, March 26, 2019. www.nytimes.com.

human decision-making out of the conflict. This element of drone warfare may again rewrite the rules of war. Throughout the history of warfare, a soldier has always had the option not to shoot, such as when conditions on the battlefield have convinced him or her that the enemy combatant does not pose a risk and should be spared. But if a computer selects the target, the element that gives a human the final decision on whether to fire a weapon at another human will have been removed from the equation.

In 2015 more than a thousand scientists, engineers, and AI experts signed a letter condemning the use of AI in military drones. The signatories of the letter stated their fear that AI technology could fall into the hands of terrorists, thereby unleashing massive weapons of destruction across the globe. According to the letter, such weapons

> will become [common] and cheap for all significant military powers to mass-produce. It will only be a matter of time until they appear on the black market and in the hands of terrorists [and] dictators wishing to better control their populace [and] warlords wishing to perpetrate ethnic cleansing. Autonomous weapons are ideal for tasks such as assassinations, destabilizing nations, subduing populations and selectively killing a particular ethnic group. We therefore believe that a military AI arms race would not be beneficial for humanity.[32]

Training Computers to Fly

Nevertheless, research to develop drones with AI capability is under way. The military calls an AI-equipped drone a lethal autonomous weapons system (LAWS). To carry out its mission, the computer aboard a LAWS drone needs to learn how to fly. This involves learning how to take off and land, how to bank left or right, and how to carry out other airborne maneuvers. Programmers can write software instructing the computer to enact the procedures to carry out those tasks, but no two flights will be the same. This means the computer will have to know where the runways begin and end at various drone bases, how to fly through inclement weather, and how to avoid other aircraft drifting too close to the drone. Military researchers are studying this hurdle and are developing a set of procedures, also known as an algorithm, that will enable the computer to learn the techniques of piloting an aircraft under different conditions.

Yet the computer still needs to learn how to fly. Charged with teaching the computers aboard LAWS drones how to fly will be US Air Force pilots. Under an air force program known as Skyborg, in 2020 actual pilots are expected to be flying fighter jets alongside drones equipped with AI. These drones are known as Valkyries. The pilots will take their planes through a number of maneuvers. The Valkyries will be programmed to mimic those maneuvers, storing in their computers' memories how to perform the various airborne movements. Under the computer's algorithm, the device stores all data that it receives, enabling it to literally learn as it goes. Engineers call this concept machine learning. The website Military.com reports, "The goal is to incorporate the Skyborg network into Valkyrie alongside manned fighters, so the machine can learn how to fly and even

algorithm

A process or set of instructions for making calculations or performing functions, often written into the software that drives a computer

The Valkyrie drone (pictured) can be programmed to mimic the maneuvers of fighter jets flown by human pilots alongside it. In this way, the drone learns how to perform maneuvers on its own through what is called machine learning.

train with its pilot. The drones will then be sent out alongside . . . fighters to scout enemy territory ahead of a strike, or to gather [intelligence] for the pilot in the formation."[33]

New Developments, New Rules

Military engineers solved the problems that plagued the Lightning Bug and other early drones nearly thirty years ago when the Predator was first dispatched over Bosnia and Herzegovina. Since then, drones have become important weapons in the military arsenals of the United States and other nations. Now, it appears engineers are close to revolutionizing drone warfare once again—by enabling the planes to fly themselves and open fire on enemies without the benefit of human decision-making. It seems likely, then, that the rules that have governed warfare for centuries will be continually rewritten as drone technology moves forward.

SOURCE NOTES

Introduction: Drones: The Future of Warfare
1. Quoted in W.J. Hennigan, "The Most-Wanted ISIS Leader Is Still Alive. Here's How the U.S. Is Using Drones to Hunt Him Down," *Time*, April 30, 2019. https://time.com.
2. Daniel Bruntstetter, "Drones: The Future of Warfare?," E-International Relations, April 10, 2012. www.e-ir.info.
3. Quoted in Jon Gambrell, "Rebel Drone Strikes Leave Two Saudi Oil Sites Ablaze," *Philadelphia Inquirer*, September 15, 2019, p. A6.

Chapter 1: What Is a Military Drone?
4. Quoted in Avery Plaw, Matthew S, Fricker, and Carlos R. Colon, *The Drone Debate: A Primer on the US Use of Unmanned Aircraft Outside Conventional Battlefields*. Lanham, MD: Rowman & Littlefield, 2016, p. 23.
5. Hugh Gusterson, *Drone: Remote Control Warfare*. Cambridge, MA: MIT Press, 2016, p. 3.
6. Kyle Mizokami, "Tiny Drone-Based Surveillance System Can Watch Over an Entire Small Town," *Popular Mechanics*, March 2, 2017. www.popularmechanics.com.

Chapter 2: Piloting a Drone
7. Quoted in Tim Wright, "Do Gamers Make Better Drone Operators than Pilots?," *Air & Space Magazine*, August 29, 2017. www.airspacemag.com.
8. Quoted in Wright, "Do Gamers Make Better Drone Operators than Pilots?"
9. Quoted in Pierre Bienaimé, "Here's How US Fighter Pilots Learn to Survive Under Inhuman Levels of G-Force," Business Insider, November 3, 2014. www.businessinsider.com.
10. Quote in S.L. Fuller, "A Day in the Life of a US Air Force Drone Pilot," *Avionics*, March 16, 2017. www.aviationtoday.com.

11. Quoted in Vegas Tenold, "The Untold Casualties of the Drone War," *Rolling Stone*, February 18, 2016. www.rollingstone.com.
12. Quoted in Matthew Power, "Confessions of a Drone Warrior," *GQ*, October 23, 2013. www.gq.com.
13. Quoted in Eyal Press, "The Wounds of the Drone Warrior," *New York Times Sunday Magazine*, June 17, 2018, p. 30.

Chapter 3: Military Drones on Land and Sea

14. Quoted in Todd South and Philip Athey, "The MAGTF Is No Longer Sacred: The Marine Corps Is Looking at Other Ways to Fight," *Marine Corps Times*, October 30, 2019. www.marinecorpstimes.com.
15. Quoted in David C. Walsh, "Marine Lab Is Rolling with New Robotic Vehicles," Defense Systems, May 12, 2015. https://defensesystems.com.
16. Quoted in David Axe, "The U.S. Army's Robot Tanks Could Arrive Years Early," *The Buzz* (blog), *National Interest*, October 14, 2019. https://nationalinterest.org.
17. Robert Sparrow and George R. Lucas Jr., "When Robots Rule the Waves?," in *One Nation Under Drones*, ed. John E. Jackson. Annapolis, MD: Naval Institute Press, 2018, pp. 78–79.
18. Quoted in Louise Knapp, "Sea Mines: An Explosive Problem," *Wired*, May 7, 2001. www.wired.com.
19. Arthur Holland Michel, "State of the Operational Art: Maritime Systems," in *One Nation Under Drones*, ed. John E. Jackson. Annapolis, MD: Naval Institute Press, 2018, p. 58.
20. Kyle Mizokami, "The U.S. Navy Wants to Build the World's Largest Robot Warship," *Popular Mechanics,* August 16, 2019. www.popularmechanics.com.

Chapter 4: Using Drones for Medical Evacuations

21. David Axe, "Army Eyes Robot Rescue Copters for Wounded Troops," *Wired,* August 2, 2012. www.wired.com.
22. Joseph Trevithick, "This Crazy Tandem Rotor Drone May One Day Rescue US Army Casualties," *War* Zone (blog), The Drive, March 28, 2017. www.thedrive.com.

23. Trevithick, "This Crazy Tandem Rotor Drone May One Day Rescue US Army Casualties."
24. Jack Stewart, "Israel's Self-Flying 'Cormorant' Whisks Wounded Soldiers to Safety," *Wired,* May 26, 2018. www.wired.com.
25. Ford Burkhart, "Robot Named Sally Steals the Show," Optics.org, May 18, 2014. https://optics.org.

Chapter 5: How Drones Are Rewriting the Rules of War

26. F.S. Naiden, "Do Drones Contradict Lessons from Centuries of War?" *Wilson Quarterly*, Fall 2013. www.wilsonquarterly.com.
27. Naiden, "Do Drones Contradict Lessons from Centuries of War?"
28. Quoted in Will Bunch, "50 Years Ago, Walter Cronkite Told America the Bitter Truth. We Need More Cronkites Today," *Philadelphia Inquirer*, February 22, 2018. www.inquirer.com.
29. Todd Gitlin, foreword to *The War Within: America's Battle over Vietnam*, by Tom Wells. Berkeley: University of California Press, 1994, p. xiv.
30. Amy Zegart, "Not All Drones Are Created Equal," *The Atlantic*, June 21, 2019. www.theatlantic.com.
31. Jacquelyn Schneider and Julia Macdonald, "US Public Support for Drone Strikes: Why Do Americans Prefer Unmanned over Manned Platforms?," Center for a New American Security, September 20, 2016. www.cnas.org.
32. *Berkeley Engineer*, "Open Letter on AI," November 1, 2015. https://engineering.berkeley.edu.
33. Oriana Pawlyk, "Air Force's Future Stealthy Combat Drone Could Use AI to Learn," Military.com, June 20, 2019. www.military.com.

FOR FURTHER RESEARCH

Books

Christopher J. Fuller, *See It/Shoot It: The Secret History of the CIA's Lethal Drone Program*. New Haven, CT: Yale University Press, 2017.

John E. Jackson, ed., *One Nation Under Drones*. Annapolis, MD: Naval Institute Press, 2018.

Kimberly Pucci, *Prince of Drones: The Reginald Denny Story*. Orlando: BearManor Media, 2019.

Paul Scharre, *Army of None: Autonomous Weapons and the Future of War*. New York: W.W. Norton, 2019.

Brett Velicovich and Christopher S. Stewart, *Drone Warrior: An Elite Soldier's Inside Account of the Hunt for America's Most Dangerous Enemies*. New York: HarperCollins, 2017.

Internet Sources

Kyle Mizokami, "Tiny Drone-Based Surveillance System Can Watch Over an Entire Small Town," *Popular Mechanics*, March 2, 2017. www.popularmechanics.com.

F.S. Naiden, "Do Drones Contradict Lessons from Centuries of War?," *Wilson Quarterly*, Fall 2013. www.wilsonquarterly.com.

Joseph Trevithick, "This Crazy Tandem Rotor Drone May One Day Rescue US Army Casualties," *War* Zone (blog), The Drive, March 28, 2017. www.thedrive.com.

Tim Wright, "Do Gamers Make Better Drone Operators than Pilots?," *Air & Space Magazine*, August 29, 2017. www.airspacemag.com.

Amy Zegart, "Not All Drones Are Created Equal," *The Atlantic*, June 21, 2019. www.theatlantic.com.

Websites

Center for a New American Security (http://drones.cnas.org). Based in Washington, DC, this group studies a number of national security issues, including the deployment of military drones. It maintains a website, Proliferated Drones, in which a number of issues are examined, including essays on how drones have rewritten the rules of war, what the American public thinks of drones, and an overview of the technology of drones.

Center for the Study of the Drone (https://dronecenter.bard.edu). Based at Bard College in Annandale-on-Hudson, New York, the organization studies issues related to drone warfare. Visitors to the group's website can download copies of *The Drone Databook*, an annual report that assesses how the militaries of ninety countries make use of unmanned aerial vehicles.

Military.com (www.military.com/equipment/drones). This website's "Drones" section provides an overview of several drones in the service of the US military, ranging from handheld reconnaissance drones to missile-carrying Predators and Reapers. Visitors can find photographs of the various drones as well as descriptions of the technology they employ to carry out their missions.

Out of the Shadows (www.outoftheshadowsreport.com). Visitors to this website can find the report by Columbia Law School's Human Rights Clinic on how the American military undercounts civilian deaths caused by drones. Visitors can also find links to numerous stories in the international media reporting on civilian deaths caused by drones.

INDEX

Note: Boldface page numbers indicate illustrations.

Abrams tank. *See* M-1 Abrams tank
Advanced Explosive Ordnance Disposal Robotics System, 40
Afghanistan
 medevac helicopter in, **39**
 UAV warfare in, 10, 23–24
algorithm, definition of, 55
American Homefront Project, 38
Anisimov, Andrei, 34
Armed Forces Health Surveillance Center, US, 25
artificial intelligence (AI)
 technology, 51–54
 definition of, 50
Axe, David, 38

bin Laden, Osama, 16
Black Hawk helicopter, 41
Black Hornet (reconnaissance drone), 13
Boucher, Max, 7–8, 11
Bruntstetter, Daniel, 5
Bryant, Brandon, 23–24
Burkhart, Ford, 45

Centeno, David, 30
Center for a New American Security, 50, 61
Center for the Study of the Drone (Bard College), 6, 61
centrifuge, definition of, 20
civilian deaths, US military's acknowledgment of, 51
conspicuous, definition of, 39
Cormorant (medevac drone), 41–44, **42**
Cronkite, Walter, 48

Denny, Reginald, 8
DP-14 Hawk (medevac drone), 40–41, 44
draft, definition of, 48
drones. *See* land-based drones; seaborne drones; unmanned aerial vehicles

Fields, John, 18
Freedberg, Sydney, Jr., 30–31

Global Hawk, **15,** 16
Gusterson, Hugh, 14

Haas, Michael, 23
hand-eye coordination, definition of, 18
Hawk. *See* DP-14 Hawk
Hellfire missiles, 5, **11,** 15, 37, 47
Houthis (Yemeni rebel group), 6
Human Rights Clinic (Columbia Law School), 51

hyperspatial/hyperspatial sensor, definition of, 12

infrared, definition of, 12
Islamic State in Iraq and Syria (ISIS), 4–5

Kennedy, Joseph, 10
Kim, Shane, 43
Klein, James, 22–23

land-based drones, 29–31
   Russian program developing, 34
large unmanned surface vehicles (LUSVs), 35–36
Leggett, Sonny, 51
lethal autonomous weapons system (LAWS), 54–56
Lightning Bug (surveillance drone), 8–9, 11–12
Lucas, George R., Jr., 31

Macdonald, Julia, 50
Marine Air-Ground Task Force (MAGTF), 27
Marine Corps Warfighting Lab, 27–28
Mead, Corey, 22
medevac/medevac helicopters, 38–39
   definition of, 38, **39**
   use of drones for, 40–42, **42**
Michel, Arthur Holland, 33–34
Military.com, 55–56, 61
mines, disarming of, **32**, 32–34
Mizokami, Kyle, 13, 15
modular, definition of, 28
Modular Advanced Armed Robotic System (MAARS), 28

M-1 Abrams tank, 30
Moseley, T. Michael, 10–11
MQ-9 Reaper, 4–5, **6**
   first use of, 10–11
   training of pilots for, **19**

Naiden, F.S., 46–47

obsolete, definition of, 5
Ogle, Alan, 25
Out of the Shadows (website), 61

Peck, Michael, 40
Pew Research Center, 50
pilots, **9**
   boredom as problem for, 22–23
   of land-based drones, 30
   post-traumatic stress disorder among, 24–26
   skill sets needed by, 17–18
   training of, 18–21, **19, 21,** 22
Piniero, James, 28
Popular Mechanics (magazine), 36
post-traumatic stress disorder (PTSD), 24–26
Predator, in peacekeeping mission in Bosnia-Herzegovina, 10
Project Aphrodite, 10

Radioplane, 8
radio transmission, 11–12
Reaper. *See* MQ-9 Reaper
remote combat vehicle (RCV), definition of, 29
Remote Environmental Monitoring Units (REMUS), 33–34

Robo Sally (medevac drone), 44–45, **45**
rotor, definition of, 41
RP-1 (monoplane), 8
Russia, land-based drone program of, 34

Sarie, Yahia, 6
Schneider, Jacquelyn, 50
Schnoor, Tim, 33
seaborne drones
  submersible, 31–34
  surface, 34–36, **35**
Sea Hunter (surface drone), 34–35, **35**
sensory devices
  artificial intelligence and, 52–53
  on self-driving car, **52**
  on unmanned aerial vehicles, 12–14
Sentinel drone, 16
Shadow, 15–16
Skyborg program, 55–56
Sparrow, Robert, 31
Stewart, Jack, 20, 43
strata, definition of, 47
supersonic, definition of, 19

Tesla, Nikola, 11
Trevithick, Joseph, 39, 41

UAV warfare
  in Afghanistan, 10, 23–24
  in fight against ISIS, 4–5
  implications for rules of warfare, 46–47
  public support for, 50

public support of wars and, 47–49
  in Vietnam War, 8–9
  in World War II, 8, 10
unmanned aerial vehicles (UAVs)
  army's land-based, 29–31
  in delivery of medical supplies, 43
  first military, 7–8
  for medical evacuation, 40–45, **42, 45**
  numbers in US inventory, 5–6
  radio transmission and, 11–12
  reasons seen as future of warfare, 5
  seaborne, 31–36, **35**
  sensory devices on, 12–14
  for surveillance, **15,** 15–16
Uran-9 (Russian land-based drone), 34

Valkyrie drone, **55,** 55–56
vertical takeoff and landing (VTOL) drones, 37–38
Vietnam War, 8–9, 48–49
  protest against, **49**
Voisin (military biplane), 7–8

Watson, Benjamin, 27
Wheatcroft, Jacqueline, 18
wide-area motion imagery (WAMI), 14–15
  definition of, 14

Zegart, Amy, 49
Zipline, 43